RAND | NATIONAL DEFENSE RESEARCH INSTITUTE

EXPEDITIONARY
CIVILIANS

Creating a Viable Practice of Civilian Deployment Within the U.S. Interagency Community and Among Foreign Defense Organizations

Molly Dunigan, Michael Schwille, Susanne Sondergaard, Susan S. Everingham, Todd Nichols

Prepared for the Office of the Secretary of Defense

For more information on this publication, visit www.rand.org/t/rr1249

Library of Congress Cataloging-in-Publication Data
ISBN: 978-0-8330-9269-4

Published by the RAND Corporation, Santa Monica, Calif.
© Copyright 2016 RAND Corporation
RAND® is a registered trademark.

Support RAND
Make a tax-deductible charitable contribution at
www.rand.org/giving/contribute

www.rand.org

Preface

Civilians now routinely deploy to support military missions abroad. Defense departments have been drawing on internal civilian capabilities to relieve pressure on the uniformed military, with some of these initiatives being formalized into organizational structures. The RAND Corporation conducted a study for the U.S. Department of Defense (DoD) that constituted an end-to-end review of guidance across the civilian deployment process, with the ultimate aim of recommending guidelines for establishing and maintaining a civilian deployment capability for the future. As part of that research, the RAND study team investigated a number of deployment approaches taken by organizations analogous to DoD, both domestic and foreign. This report presents the findings from that effort.

This research was conducted in 2014, and the findings were current as of mid-2015. It should be of interest to decisionmakers in country-level and international organizations who are tasked with developing strategies for civilian deployment and to military officials in both the United States and abroad who issue requirements for deployed civilians, work alongside deployed civilians, or are interested in the issue of civilian deployment. This report should also prove useful to researchers and policymakers who are interested in workforce mix issues or developing methods and strategies to more effectively and efficiently use civilians to fulfill critical roles in deployed settings.

A companion to this report, *Expeditionary Civilians: Creating a Viable Practice of Department of Defense Civilian Deployment*, provides the full study results and is available at www.rand.org/t/RR975.

This research was sponsored by the Office of the Deputy Assistant Secretary of Defense for Civilian Personnel Policy and conducted within the Forces and Resources Policy Center of the RAND National Defense Research Institute, a federally funded research and development center sponsored by the Office of the Secretary of Defense, the Joint Staff, the Unified Combatant Commands, the Navy, the Marine Corps, the defense agencies, and the defense Intelligence Community.

For more information on the RAND Forces and Resources Policy Center, see www.rand.org/nsrd/ndri/centers/frp or contact the director (contact information is provided on the web page).

Contents

Figures and Tables

Figures

Tables

Summary

Civilians now routinely deploy to support military missions abroad. Internationally, defense departments have been drawing on internal civilian capabilities to relieve pressure on the uniformed military, with some of these initiatives being formalized into organizational structures.

There are several known challenges associated with deploying civilians to operational theaters. For instance, from where should the capability be drawn? How should deployable civilians be selected, prepared, and protected in theater? How can an organization best manage civilians while they are deployed, ensuring that they will have secure jobs upon their return? Moreover, from a recruitment standpoint, how can an organization ensure a steady pipeline of willing volunteers to deploy? How are civilians perceived by and how do they operate among their military colleagues? These are challenges that organizations attempting to deploy civilians will need to address.

The RAND Corporation conducted a study for the U.S. Department of Defense (DoD) that constituted an end-to-end review of guidance across the civilian deployment process, with the ultimate aim of recommending guidelines for establishing and maintaining a civilian deployment capability for the future. The results of that study can be found in the companion report, *Expeditionary Civilians: Creating a Viable Practice of Department of Defense Civilian Deployment*, available at www.rand.org/t/RR975. As part of the research for that larger report, we investigated a number of deployment approaches taken by organizations analogous to DoD, both U.S. and foreign. This report presents the specific findings derived from that aspect of the larger research project. The research was conducted in 2014, and the findings were current as of mid-2015.

From our comparative cases, we identified best practices and created a typology of four models of civilian deployment, highlighting the benefits and drawbacks of each. We then developed an overall assessment of the viability of the current civilian deployment concept and devised recommendations for establishing and maintaining a civilian deployment capability that could feasibly meet the requirements for an expeditionary civilian capability over the next several decades.

To support this research effort, we conducted interviews with representatives from government agencies in the United States, Canada, the United Kingdom, and Australia with well-established civilian deployment programs to learn more about the requirements that generate the need for deployable civilians, the types of missions civilians support, and the methods that organizations use to identify, select, track, and deploy civilians.

We found that each organization interviewed had unique missions and challenges; as a result, they used a variety of methods to deploy their personnel. While some organizations had a narrowly focused mission set, others were responsible for a wide-ranging set of missions.

Typology of Civilian Deployment Models

We identified four models that organizations have applied to deploy civilians. The models differ along two main dimensions: the extent to which they sourced individuals to deploy from within the organization's existing civilian ranks (internal sourcing), as opposed to searching for candidates external to the organization (external sourcing), and the extent to which the organizations had a pool of preidentified individuals prior to the issuance of requirements (proactive sourcing), as opposed to identifying candidates for positions after requirements had been issued (reactive sourcing).[1]

---

[1]   There are benefits and drawbacks to both reactive and proactive sourcing, depending on the situation in question.

We defined the four models as follows:

- *Reactive internal sourcing.* A requirement is identified through either a top-down or bottom-up process, triggering a recruitment process internal to the organization. Individuals are selected to fill the requirement, after which they undertake any required pre-deployment training or preparation (e.g., medical screening and vaccinations, visas, clearances, culture training, hostile environment training) as needed for the specific deployment. Individuals deploy to their posting, and after deployment they return to the post they had occupied prior to deployment.
- *Proactive internal sourcing.* Deployed civilians are sourced from within an organization's existing civilian employee pool. Rather than waiting for a specific requirement to be identified, the most probable requirements to emerge are preidentified. Civilians can then apply to be part of a readiness pool that will be used to source the set of emergent requirements.
- *Reactive external sourcing.* A specific requirement is identified, and the organization advertises the requirement externally. An outside expert is then hired to fill the specific requirement. Following deployment, employees hired under this model are no longer affiliated with the organization, often returning to their former posts with other organizations.
- *Proactive external sourcing.* Personnel are identified from outside an organization to fill requirements. A set of future requirements is forecasted using various planning models, and then personnel are hired to source those requirements. In anticipation of a requirement for civilian deployment, organizations set up a bullpen, a readiness pool of external selectees that is used to fill requirements when needed. When a requirement is issued that matches the qualifications of a particular individual in the bullpen, that individual is notified for a deployment and then his or her pay and benefits are activated.

We believe that organizations can draw on a combination of these models to best respond to requirements. We found it notable that, of the four models we identified, those that involved a proactive sourcing approach allowed organizations to deploy personnel significantly faster than those that involved recruiting qualified personnel after a requirement had been issued. Meanwhile, organizations that relied on external sourcing models spoke to the numerous rules and regulations faced by government agencies. At times, lengthy justifications were needed to select one individual over another to adhere to fair hiring practices. This indicates that, when reactive sourcing is necessary, sourcing officials may need direct or expedited hiring authorities to enhance their capabilities to source positions quickly. Furthermore, regardless of whether individuals were sourced internally or externally, some type of oversight organization was necessary to ensure the successful deployment of civilians.

Conclusions and Recommendations

This research points to a number of interesting findings regarding the benefit of a long-term strategy aimed at developing a viable civilian deployment practice that will be sustainable and the specific practices and processes that organizations might usefully employ on a day-to-day basis to ensure effective, efficient civilian deployments. It is useful to consider the management of civilian deployment capabilities as being divided into three categories of activities: policy, planning and strategy, and operations. *Policy* responsibilities entail writing policy to determine the guidelines for civilian deployments. *Planning and strategy* responsibilities entail mission-based, scenario-specific forecasting and strategic human capital planning. *Operational* responsibilities entail the assignment of requirements, sourcing, readiness preparation, and during/postdeployment tracking of expeditionary civilians.

Policy
Championing Expeditionary Civilian Capabilities
The military may be unaware of the benefits of drawing on civilian capabilities, and organizations must make a conscious effort to market their capabilities. This is a relatively day-to-day practice that organiza-

tions can employ to ensure that defense leaders and other "customers" of deployable civilian capabilities are aware of the benefits that deployed civilians can bring to an operation.

Balancing Readiness with Cost

The speed of recruitment and the cost of readiness preparation and maintenance must be balanced against the cost of those capabilities. Those capabilities vary to different degrees under different circumstances, and some skills should likely be maintained internally over the long term (while others need not be). However, there is not one model that is right for every organization, and each should weigh its own requirements to deploy civilians and adopt an appropriate approach.

Planning and Strategy
Speed Versus Capability

Our analysis suggests that there is a need to examine the speed of the recruitment process, the cost of maintaining deployable civilians at a given level of readiness, and the possibility of developing a preselected pool or making deployment part of the job description. Furthermore, we found that if a specific skill set is required within an organization for the future, it is valuable to develop and sustain the capability internally rather than externally to be able to deploy individuals with that capability in the future.

Planning and Forecasting

A failure to effectively integrate expeditionary civilians into planning hinders the development of realistic expectations and poses a challenge to the integration of these civilians. Therefore, organizations need to dedicate resources to the planning and forecasting of future requirements to optimize the organizational design associated with civilian deployments.

Operations
Centralized Versus Decentralized

Organizations must determine the level at which to manage many of the processes that govern civilian deployments. The extent to which an organization centralizes management of civilian deployment processes determines the speed, budget, capability, and resources that it requires. A decentralized recruitment, screening, and selection approach does not necessarily work when an organization must meet several different deployable position requirements. A centralized process may be slow to respond to emergent requirements.

Sharing Resources

We identified opportunities to pool and share existing enabling capabilities for civilian deployments, including the sharing of training facilities. Consolidating the responsibility to deploy personnel from multiple organizations to one deployment center saves on overhead, personnel, and operational costs.

Areas for Future Research

We identify several potential routes for future research. First, the development of a transparent costing framework would be quite useful in supplementing the current dearth of cost analysis on different deployment models or, indeed, the costs associated with different utilization of parts of the whole force. Second, future research could also usefully examine, compare, and analyze various methods for forecasting deployable civilian requirements. Third, future research could focus on the practical aspects of civilian deployment, such as safety considerations, operating with military personnel, performance metrics, and the psychological impact of deployments. Finally, research could be conducted to further elaborate and refine the deployment models.

Acknowledgments

We gratefully acknowledge the assistance of a number of individuals across the U.S. Department of Defense, Joint Staff, combatant commands, U.S. military services, other U.S. government agencies, and foreign governments who took the time to speak with us for this study. Although we cannot name them publicly, we are extremely grateful for their assistance. At RAND, we thank John Winkler, Lisa Harrington, and Jennifer Lamping-Lewis for their management support and Lauren Skrabala for her writing and editorial support. We are also very grateful to Ginger Groeber, Wade Markel, and Agnes Schaefer for their careful review of the parent report from which this research is derived (*Expeditionary Civilians: Creating a Viable Practice of Department of Defense Civilian Deployment*). We extend special thanks to the Office of the Deputy Assistant Secretary of Defense for Civilian Personnel Policy, which funded this research, and to Deputy Assistant Secretary Paige Hinkle-Bowles, Julie Blanks, and Joe Daniel for the guidance they provided over the course of the study.

Abbreviations

CCMD	combatant command
COM	chief of mission
CPCC	Civilian Planning and Conduct Capability
CS3	Crisis Surge Support Staff
DoD	U.S. Department of Defense
DoS	U.S. Department of State
DSTO	Australian Defence Science and Technology Organisation [now the Defence Science and Technology Group]
EEAS	European External Action Service
FEMA	Federal Emergency Management Agency
HR	human resources
MOD	UK Ministry of Defence
NATO	North Atlantic Treaty Organization
S2O	Support to Operations
USAID	U.S. Agency for International Development
U.S.C.	U.S. Code

Expeditionary Civilians: Creating a Viable Practice of Civilian Deployment Within the U.S. Interagency Community and Among Foreign Defense Organizations

Introduction

Background

As defense and interagency organizations move toward even slimmer workforces with less capability redundancy, there is a need to look at alternative and sustainable approaches to maintaining access to resources. For some time, defense departments have drawn on internal civilian capabilities to relieve pressure on the uniformed military, with some of these initiatives being formalized into organizational structures. In 2007, civilian leadership in the U.S. Office of the Secretary of Defense determined that the approximately 700,000-person civilian workforce in the U.S. Department of Defense (DoD) could be a viable source of deployable personnel. In 2009, this arrangement was formalized with the establishment of the Civilian Expeditionary Workforce by DoD Directive 1404.10. Around the same time, the UK Ministry of Defence (MOD) set up the Support to Operations (S2O) office to provide sustainable civilian support to operations. Meanwhile, such defense-sector reliance on civilian capabilities increased the need for interagency organizations—such as the U.S. Department of State (DoS), U.S. Department of Homeland Security, and the U.S. Agency for International Development (USAID)—to deploy their personnel in support of defense missions.

There are several known challenges associated with the deployment of civilian capability to operational theaters. For instance, from where should the capability be drawn? How should deployable civilians be selected, prepared, and protected in theater? How can an organization best manage civilians while they are deployed, ensuring that they will have secure jobs upon their return? Moreover, from a recruitment standpoint, how can an organization ensure a steady pipeline of willing volunteers to deploy? How are civilians perceived by and how do they operate among their military colleagues? These are challenges that organizations attempting to deploy civilians will need to address.

RAND conducted research for DoD constituting an end-to-end review of guidance across the civilian deployment process, with the ultimate aim of recommending guidelines for establishing and maintaining a civilian deployment capability for the future. As part of that project, we investigated a number of deployment approaches taken by organizations analogous to DoD, both domestic and foreign. The study identified lessons and promising practices for civilian deployment that could be applied, in general terms, to a broad spectrum of organizations seeking to deploy civilians. The results of that study are available in the companion report, *Expeditionary Civilians: Creating a Viable Practice of Department of Defense Civilian Deployment*.[1] This report is derived from that larger RAND study, but it focuses solely on the findings from the analysis of alternative civilian deployment models used by non-DoD interagency and by defense organizations in the United States and internationally. The goal of this report is to provide information on the civilian deployment practices and procedures of the North Atlantic Treaty Organization (NATO) and its individual members' defense organizations. This research was conducted in 2014, and the findings were current as of mid-2015.

Structure of This Report

We begin by presenting the methodology used to analyze several distinct cases of civilian deployment across both U.S. and international interagency and defense organizations. We then describe key terms and concepts that are common throughout organizations that deploy civilian personnel, with a focus on authorities, requirements, and mission

---

[1]   Molly Dunigan, Susan S. Everingham, Todd Nichols, Michael Schwille, and Susanne Sondergaard, *Expeditionary Civilians: Creating a Viable Practice of Department of Defense Civilian Deployment*, Santa Monica, Calif.: RAND Corporation, RR-975-OSD, 2016.

types. We provide short descriptions of each organization for context. We then present a typology of four deployment models, highlighting the advantages and disadvantages associated with each. The final sections of the report present lessons identified from the research, conclusions, and recommendations.

Methodology and Cases

For the larger study from which this report is derived, we worked with DoD officials to identify the goals of DoD's civilian deployment capability and potential policy and planning gaps. We then reviewed U.S. combatant command (CCMD) requirements for expeditionary civilians, as well as the CCMDs' various policies and practices regarding the deployment of civilians.[2] The findings established baseline manpower and personnel management requirements for future requirements to support contingency program management over time. We then drew on relevant lessons from comparative cases of civilian deployment policies and practices from both domestic U.S. government organizations and foreign governments. From these comparative cases, we identified best practices and created a typology of four models of civilian deployment, highlighting the benefits and drawbacks of each. Combining the lessons from this typology with the policy analysis and survey of CCMD needs for expeditionary civilian capabilities, we developed an overall assessment of the viability of the current civilian deployment concept and devised recommendations for establishing and maintain-

---

[2] A combatant command is defined in Joint Publication 1-02 as follows: "A unified or specified command with a broad continuing mission under a single commander established and so designated by the President, through the Secretary of Defense and with the advice and assistance of the Chairman of the Joint Chiefs of Staff" (U.S. Joint Chiefs of Staff, *Department of Defense Dictionary of Military and Associated Terms*, Joint Publication 1-02, Washington, D.C., as amended through November 15, 2015). There are six geographical commands; U.S. Africa Command, U.S. Central Command, U.S. European Command, U.S. Northern Command, U.S. Pacific Command, and U.S. Southern Command. There are also three functional commands: U.S. Special Operations Command, U.S. Strategic Command, and U.S. Transportation Command.

ing a civilian deployment capability that could feasibly meet CCMD requirements over the next several decades.

Much of the data for these tasks were collected through interviews with DoD and CCMD officials, as well as a review of relevant policy guidance, analysis and assessments by such organizations as the U.S. Government Accountability Office, and secondary literature. Over the course of the study, we interviewed a total of 83 individuals spanning 45 offices across DoD, other U.S. government agencies, and foreign governments. Interviewees included representatives of multiple directorates under the Office of the Secretary of Defense, the military services, and each of the geographic CCMDs (U.S. Central Command, U.S. Southern Command, U.S. European Command, U.S. Africa Command, U.S. Pacific Command, and U.S. Northern Command), in addition to one functional CCMD (U.S. Special Operations Command).[3]

To identify best practices from comparative cases we first conducted a short literature review that provided a general understanding of civilian deployments beyond DoD practices, the requirements that feed those deployments, and how those requirements are sourced. After establishing a baseline understanding of how organizations other than DoD deploy their personnel, we developed a list of U.S. and foreign government agencies that deploy civilians to at least some extent. In selecting cases for inclusion in the analysis, we sought variation in terms of the length of agencies' experiences with civilian deployment practices, the numbers of civilians typically deployed, and the purposes for which civilians are deployed. One of the foremost goals in selecting cases for analysis was that the universe of cases analyzed should reflect organizations similar to DoD in at least one of these respects.

---

[3]   These interviews are attributed anonymously throughout this report in compliance with the U.S. Federal Policy for the Protection of Human Subjects (also known as the Common Rule). Both RAND's Institutional Review Board and human-subjects protection reviewers in DoD approved of this research method for this study. Organizational affiliation is included in the citation for each anonymous interviewee to give a sense of the individual's background and experience, but it should be noted that interviewees were not asked to represent their organizations in a confidential way. While interviewees were asked to respond based on their professional experiences, they were, in all cases, speaking for themselves rather than for their organizations in an official capacity.

As a result of this case selection method, our interviews included discussions with representatives of several DoD Fourth Estate agencies that have their own well-established civilian deployment programs,[4] such as the Defense Logistics Agency, Defense Intelligence Agency, and the Defense Contract Management Agency.[5] Other U.S. interagency organizations with which we conferred included the U.S. Department of State's Bureau of Conflict and Stabilization Operations and Bureau of Diplomatic Security; the U.S. Department of Homeland Security's Office of International Affairs, U.S. Customs and Border Protection, the Drug Enforcement Administration, and the Federal Emergency Management Agency (FEMA); and USAID's Office of Crisis Surge Support Staff (CS3) and Office of Transition Initiatives. Outside of the United States, we spoke with officials from MOD's S2O office; the Canadian Department of National Defence; the Australian Defence Science and Technology Organisation (DSTO); and the European Union's European External Action Service (EEAS) Civilian Planning and Conduct Capability (CPCC), including a representative from CPCC's Civilian Response Teams.

In conducting the research, we found that each organization interviewed had unique missions and challenges; as a result, they used a variety of methods to deploy their personnel. While some organizations had a narrowly focused mission set, others were responsible for a wide-ranging set of missions. To accurately reflect this variation, we ultimately decided to interview a set of organizations representing a diverse workforce covering a variety of missions.

Table 1 lists the organizations and types of personnel with whom we conducted interviews. Our data collection sample consisted of

---

[4] The "Fourth Estate" is all of the organizational entities in DoD that are not in the military departments (services) or the CCMDs. They include the Office of the Secretary of Defense, the U.S. Joint Chiefs of Staff, DoD's Office of the Inspector General, the defense agencies, and DoD field activities.

[5] Although the focus of this segment of the research was on non-DoD organizations, we included these Fourth Estate agencies in the analysis because each has its own civilian deployment process that is distinct from those of DoD writ large. Therefore, we sought to determine the extent to which these distinct processes employed best practices with the potential to usefully inform other organizations' approaches to civilian deployment.

Table 1
Organizations Interviewed

Agency Type		Number of Interviewees
U.S. government agencies		
DoS	Bureau of Conflict and Stabilization Operations	1
	Bureau of South and Central Asian Affairs	1
	Bureau of Diplomatic Security	1
	Afghanistan and Pakistan Strategic Partnership Office	1
USAID	CS3	1
	Office of Transition Initiatives	1
U.S. Department of Homeland Security	Office of International Affairs	1
	U.S. Customs and Border Protection	2
	FEMA	1
U.S. Department of Justice	Drug Enforcement Administration	2
DoD Fourth Estate	Defense Intelligence Agency	5
	Defense Logistics Agency	1
	Defense Contract Management Agency	3
Foreign government agencies		
UK MOD	S2O	2
Canadian Department of National Defence	J1 (personnel) and human resources	2
Australian Department of Defence	DSTO	1
EEAS	CPCC	6
	CPCC Civilian Response Teams	1
Total		33

interviews with 33 personnel from government agencies both inside and outside the United States.

Throughout the interview process, we accumulated a wealth of knowledge regarding specific civilian deployment experiences including the requirements that generate the need for deployable civilians, the types of missions civilians support, and the methods that organizations use to identify, select, track, and deploy civilians. These findings and promising practices are outlined in subsequent sections of this report.

Overview of Case Characteristics

There are three main concepts that both characterize and influence the deployment approach used by organizations: The authority (or authorities) under which civilians deploy, the source and content of the requirements for resources, and the type of missions that civilians are sourced to support. In the following sections, we define each of the concepts as applied within the context of this study.

Defining Key Terms and Concepts
Authorities

Civilians routinely deploy to support missions through a variety of authorities. The U.S. government agencies included in our analysis deploy civilians who typically operate either under Chief of Mission (COM) authorities derived from Title 22 of the U.S. Code (U.S.C.) or authorities derived from Title 10 U.S.C. through DoD.[6] Under most circumstances, deployed personnel are ultimately the responsibility of either the U.S. ambassador or a military commander.

Most civilian U.S. agencies deploy their personnel to a contingency operation under COM authority. The DoS *Foreign Affairs Manual, Volume 2*, clearly describes COM authority and the processes

---

[6] Title 22 outlines the role of foreign relations and intercourse. It is broken down into more than 86 chapters that cover a wide range of authorities and activities conducted by the U.S. government. Title 10 outlines the role of armed forces and provides the legal basis for the roles, missions and organization of each of the uniformed services (Air Force, Army, Marine Corps, and Navy) and DoD.

to exercise that authority over U.S. government staff and personnel for missions abroad:

> COMs are the principal officers in charge of U.S. Diplomatic Missions and certain U.S. offices abroad that the Secretary of State designates as diplomatic in nature. The U.S. Ambassador to a foreign country, or the *chargé d'affaires*, is the COM in that country.[7]

A number of documents provide guidance and the legal basis for these authorities, including the President's letter of instruction to COMs, the DoS Basic Authorities Act, the 1980 Foreign Service Act, the 1986 Diplomatic Security Act, and National Security Decision Directive 38.[8] The COM has authority over every executive-branch agency in a host country, with the exception of personnel under the command of a U.S. military commander—typically the combatant commander or geographic combatant commander—and personnel on the staff of an international organization.

The other types of relevant authorities are derived from Title 10 U.S.C. and are inherently military in nature. In selecting forces for various missions, combatant commanders consult such strategic guidance documents as the *Unified Command Plan, National Security Strategy, National Defense Strategy, National Military Strategy,* the *Quadrennial Defense Review, Guidance for the Development of the Force, Guidance for Employment of the Force,* and *Joint Strategic Capabilities Plan.*[9] The combatant commander then requests forces through the U.S. Joint Chiefs of Staff, which, in turn, makes recommendations to the Secre-

---

[7] U.S. Department of State, *Post Management Organization, Chief of Mission Authority and Overseas Staffing,* Foreign Affairs Manual 2 FAH-2, H-110, July 18, 2014, p. 2.

[8] For a full review of these documents, see U.S. Department of State, 2014.

[9] See, for example, Office of the President of the United States, *National Security Strategy,* Washington, D.C., May 2010; U.S. Department of Defense, *Sustaining U.S. Global Leadership: Priorities for 21st Century Defense,* Washington, D.C., January 2012; U.S. Joint Chiefs of Staff, *The National Military Strategy of the United States of America: Redefining America's Military Leadership,* Washington, D.C., February 2011; U.S. Department of Defense, *Quadrennial Defense Review Report,* Washington, D.C., February 6, 2006; and U.S. Department of Defense, *Quadrennial Defense Review Report 2014,* Washington, D.C., March 4, 2014.

tary of Defense. The Secretary of Defense then assigns military personnel and civilians to CCMDs for mission execution. Typically, civilians are employed under one of these authorities when deployed in support of a contingency operation.

Due to the increased terrorism threat and the need to ensure security responsibility for DoD personnel and facilities in foreign areas, the Secretary of State and the Secretary of Defense signed a memorandum of understanding with the effective date of December 16, 1997, to identify security responsibilities.[10] Pursuant to this memorandum, responsibility for security has been clearly delineated between the two departments through a series of individual country agreements, via an individual memorandum of agreement, assigning responsibility for the security of DoD personnel in a given country to either the COM or the combatant commander, depending on the mission. This agreement has served to clarify previous confusion over the security responsibilities of DoD personnel.[11] The COM authority and other Title 10 authorities cover U.S. personnel only. As noted, the COM authority typically covers civilian agencies, and the Title 10 authorities cover personnel assigned to CCMDs.

Other countries have different rules and regulations that govern the employment of civilians. However, while each country has its own legal rules and regulations that must be satisfied to deploy personnel, representatives from most of the organizations consulted for this study did report having a similar framework under which civilians deploy. Similarly, in the context of European Union Common Security and Defence Policy civilian missions, civilians deploy under an EU mandate approved by the Council of the European Union.

Requirements

Requirements drive operational missions. For the purposes of this study, we viewed requirements as the set of activities necessary to develop,

---

[10] "Memorandum of Agreement Between the Departments of State and Defense on the Protection and Evacuation of U.S. Citizens and Nationals and Designated Other Persons from Threatened Areas Overseas," September 1997.

[11] U.S. Department of State, 2014.

consolidate, coordinate, validate, approve, and prioritize the deployment of civilian personnel for contingency operations. Many of the agencies examined here have adapted portions of their organizations, business processes, and deployment models to address requirement requests. Where requirements come from greatly affects the organization and processes that an agency uses to deploy personnel. Therefore, it is critical to understand where requirements are generated and how they are processed. Typically, there are two methods through which a request can come to an agency: top-down or bottom-up.

Top-down requests in the U.S. context originate from the division, bureau, or secretariat/headquarters levels; the National Security Council; or congressional or presidential direction. Top-down requests are primarily directive in nature and compel the organization to react. Bottom-up requests, on the other hand, come to an agency from a variety of sources outside the organizational chain of command and either could require immediate attention or could be staffed through routine procedures. Here, the requesting agent often makes the request through a U.S. embassy, either on behalf of a partner nation or via the COM. Requests can also come through other federal agencies or through NATO, the European Union, the United Nations, or another international entity.

Correctly identifying the origin of the bulk of requirements will guide the type of deployment model an organization uses. Interviewees described a need to balance efficiency and speed with personnel identification. In general, the speed with which a requirement must be filled will determine whether the individual selected for a deployment should come from within the organization or whether he or she can be hired from outside the organization; it also determines whether there should be a preselected pool of candidates prior to requirement identification. The typology of organizational structures is discussed in more detail later in this report.

Mission Types

There are a variety of missions that civilian agencies routinely deploy personnel to support, ranging from relatively benign workshops and technical assistance programs to efforts aimed at countering extremist operations in high-threat environments. The agencies interviewed for this study covered a suitably diverse host of "non–steady-state" operations that require civilian expertise, including humanitarian assistance/ disaster relief, stabilization and reconstruction, counterdrug, counterpiracy, capacity building, institution building, election monitoring, intelligence, countergang, technical assistance, liaison and coordination duties, and security force training.

While military personnel can and do conduct many of these missions, in some cases, specific civilian expertise is desired. The key is discerning when to leverage civilian expertise versus when a generalist will suffice. Many of the skills required for the mission types listed above center on specific expertise that is found predominantly within the civilian workforce.

Case Summaries

In this section, we present a brief summary of each of the organizations that we approached for this study. While we mention deployment numbers, which were obtained through the interview process, it is important to note that because our focus was on the *process* of deployment, we did not comprehensively analyze and cross-check the number of personnel deployed by each organization. We did find, however, that the number of civilians deployed by a given organization tended to vary on an annual basis, depending on missions and requirements. Furthermore, not all organizations were in a position to provide us with an exact number of civilians deployed in a given time frame. We include the estimated numbers to give a rough indication of the size and scope of civilian deployments from each organization.

Table 2 captures many of the findings from our interviews. The "Deployment Type" column indicates whether requirements are part of steady-state operations or are typically emergent requests. The "Deployment Office" column indicates the structure of the office that deploys civilians; centralized offices maintain more oversight of deployed personnel, and decentralized offices relinquish more control to field offices and the individual. The "Requirement Source" column indicates where

Table 2
Characteristics of the Organizations Interviewed

Organization Type	Annual Number	Deployment Type	Deployment Office	Requirement Source	Sourcing	Volunteers
U.S. government agencies						
DoS						
Bureau of Conflict and Stabilization Operations	30	Both short-notice and planned	Decentralized office	COM, functional bureaus, CCMD	Through embassy or regional or functional bureau; identify need through the Crisis Response Network	Yes
Bureau of South and Central Asian Affairs	Declined to comment	Both short-notice and planned	Centralized office	COM, functional bureaus, CCMD	Through embassy or regional or functional bureau	—
Bureau of Diplomatic Security	Declined to comment	Both short-notice and planned	Centralized office	Embassy	Internal to the bureau	Yes
Afghanistan and Pakistan Strategic Partnership Office	50–100	Planned	Centralized office	Embassy	Hire externally for most positions	Yes
USAID						
CS3	50	Both short-notice and planned	Decentralized office	Embassy	Bullpen	Part of job
Office of Transition Initiatives	190	Both short-notice and planned	Decentralized office	Embassy	Bullpen	Part of job

Table 2—Continued

Organization Type	Annual Number	Deployment Type	Deployment Office	Requirement Source	Sourcing	Volunteers
U.S. Department of Homeland Security						
Office of International Affairs	?	Both short-notice and planned	Decentralized office	Embassy	Internally for most positions	Yes
U.S. Customs and Border Protection	750	Both short-notice and planned	Centralized office	Embassy	Volunteers and internal staffing	Yes
FEMA	4,000	Short-notice	Centralized office	National Response Coordination Center	Via executive office and a declared emergency	Yes
U.S. Department of Justice						
Drug Enforcement Administration	800 positions overseas	Both short-notice and planned	Centralized office	Embassy, long-standing offices	Lengthy internal process	Yes
DoD Fourth Estate						
Defense Intelligence Agency	100–150	Both short-notice and planned	Centralized office	CCMD, Global Force Management	Through mission managers within each directorate, internally	Part of job
Defense Logistics Agency	200–300	Both short-notice and planned	Centralized office	CCMD, Global Force Management	Volunteers and internal staffing	Yes
Defense Contract Management Agency	50–100	Planned	Centralized office	CCMD, Global Force Management, joint task force, forward-stationed contract management office	Moving to be a source provider, not an executor; the services will execute their contracts	Yes

Table 2—Continued

Organization Type	Annual Number	Deployment Type	Deployment Office	Requirement Source	Sourcing	Volunteers
Foreign government agencies						
UK MOD						
S2O	150	Both short-notice and planned	Centralized office	Theater or permanent joint headquarters or elsewhere	From MOD civil servants; sometimes the wider UK civil service	Yes
Canada						
Department of National Defence (J1)	100	Both short-notice and planned	Centralized office	Theater	Within the pool of existing public servants	Yes
Australia						
DSTO	10	Both short-notice and planned	Centralized office	Operational commanders	Within DSTO or Australian Department of Defence	Yes
EEAS						
CPCC	3,200 currently deployed	Both short-notice and planned	Centralized office	Committee for Civilian Aspects of Crisis Management	Within the member state; alternatively, third states if required	Yes
CPCC Civilian Response Teams	—	Short-notice	Centralized office	Committee for Civilian Aspects of Crisis Management	Within the member state; alternatively, third states if required	Yes

most requirements are generated. The "Sourcing" column indicates the source of personnel who fulfill deployable civilian requirements. Finally, the "Volunteers" column indicates whether a force solicits volunteers for a deployment.

Bureau of Conflict and Stabilization Operations, U.S. Department of State

The Bureau of Conflict and Stabilization Operations "advances U.S. national security by breaking cycles of violent conflict and mitigating crises in priority countries."[12] It falls under the purview of the Office of the Under Secretary of State for Civilian Security, Democracy, and Human Rights and was created by Secretary of State Hillary Clinton in 2012 to improve the U.S. response to conflicts and crises in other countries. Its missions often require civilian involvement and expertise, and it has historically deployed around 30 personnel a year.[13] The bureau has also created a pool of civilian experts with specific characteristics who can be alerted and deployed on short notice, typically within two weeks.[14]

Bureau of Diplomatic Security, U.S. Department of State

The Bureau of Diplomatic Security is responsible for security and law enforcement within DoS. Inside the United States, it is responsible for the protection of the Secretary of State and visiting high-ranking dignitaries and other visiting officials. Overseas, it provides personnel and embassy security in more than 160 foreign countries across 275 U.S. diplomatic missions. It can deploy personnel as individuals or in a variety of teams, including security support teams, tactical support teams, and mobile training teams. It leads international investigations into passport and visa fraud, conducts personnel security investiga-

---

[12] U.S. Department of State, Bureau of Conflict and Stabilization Operations, web page, undated.

[13] Interview with a Bureau of Conflict and Stabilization Operations official, September 18, 2014.

[14] Interview with a Bureau of Conflict and Stabilization Operations official, September 18, 2014.

tions, and assists in threat analysis, cyber security, and counterterrorism missions.[15]

Office of Crisis Surge Support Staff and Office of Transition Initiatives, U.S. Agency for International Development

Both CS3 and the Office of Transition Initiatives deploy U.S. personnel with the necessary skill sets to further U.S. foreign interests with the goal of improving lives and livelihoods in the developing world. One of the distinguishing characteristics of these organizations is the ability to provide a surge capability to U.S. missions through a flexible and quick-reaction deployment mechanism that selects, prescreens, trains, and holds individuals in a wait status until a requirement emerges. Individuals in this wait status are said to be "on the bench" or "in the bullpen." Each office deploys between 50 and 190 personnel annually.[16] While CS3 tends to focus directly on short-term U.S. embassy support, the Office of Transition Initiatives works primarily through implementation partners to quickly provide goods and services in crisis situations.[17]

U.S. Customs and Border Protection, U.S. Department of Homeland Security

U.S. Customs and Border Protection is charged with securing more than 7,000 miles of U.S. land borders and 328 ports of entry. It is responsible for protecting U.S. citizens from terrorist threats and preventing the illegal entry of persons and goods.[18] The agency also facilitates the lawful travel and trade of goods and services across U.S borders. It has more than 42,000 officers and border-control agents who are deployed throughout the United States. Outside the United States, more than 750 agency personnel operate under COM authority in a variety of roles, including as attachés, advisers, representatives, and security personnel in support of specific missions and programs. U.S.

---

[15] U.S. Department of State, Bureau of Diplomatic Security, web page, undated.

[16] Interviews with Office of Transition Initiatives and CS3 officials, September 15, 2014.

[17] Interviews with Office of Transition Initiatives and CS3 officials, September 15, 2014.

[18] U.S. Customs and Border Protection, "About CBP," web page, undated.

Customs and Border Protection requirements are generated through a variety of multiyear initiatives, as well as quick staffing solutions to fulfill short-term, ad hoc needs.[19] To fill these latter requirements, the agency has developed a database of prescreened personnel, centered on a core group of 22 staff who can conduct short-notice training events.[20]

Federal Emergency Management Agency, U.S. Department of Homeland Security

FEMA's primary role is to coordinate the response of federal, state, and local authorities in the event of a natural or man-made disaster. The organization has more than 23 different directorates, ten regional operations centers, and an incident management and support staff of more than 17,000 personnel.[21] It has a tiered approach to readiness that allows some disaster response experts to deploy quickly while simultaneously notifying other FEMA employees of the disaster and that they might be required to deploy. For example, the incident management staff are full-time, fully trained FEMA employees who respond immediately in the event of a disaster. Depending on the severity of an event, ancillary support personnel can be called to help augment the incident management staff. Ancillary support can come from local, state, or other directorates within FEMA. The FEMA Corps, a cadre of 18- to 24-year-olds dedicated to disaster response, is one such organization that can be used in a disaster. It consists of a small number of highly skilled disaster assistance operators and is kept in a high state of readiness to deploy on short notice.[22] The U.S. Department of Homeland Security Surge Capacity is another standby force. It consists of 4,000 federal employees who can be called in the event of an emergency to provide additional capability to FEMA.[23]

---

[19] U.S. Customs and Border Protection, undated.

[20] Interview with U.S. Customs and Border Protection officials, September 18, 2014.

[21] Interview with a FEMA official, September 19, 2014.

[22] Interview with a FEMA official, September 19, 2014.

[23] Interview with a FEMA official, September 19, 2014.

Drug Enforcement Administration, U.S. Department of Justice

The mission of the Drug Enforcement Administration is "to enforce the controlled substances laws and regulations of the United States and bring to the criminal and civil justice system of the United States, or any other competent jurisdiction, those organizations and . . . members of organizations, involved in the growing, manufacture, or distribution of controlled substances."[24] The office has roughly 800 positions overseas, representing approximately 10 percent of its workforce, and has been deploying agents and support staff to overseas missions under COM authority for more than 25 years.[25] Deployed personnel support a variety of missions and activities, including the management of a national drug intelligence program, investigation and preparation of cases for prosecution, liaison and coordination duties, training activities, and investigative and strategic intelligence gathering.

Defense Intelligence Agency, U.S. Department of Defense

The Defense Intelligence Agency has deployed a range of operational and support personnel since the Vietnam War, but it was not until 2002 that it began emphasizing the deployment of civilian personnel. Personnel routinely deployed since 2002 include analysis, intelligence collection, IT support, logistics, administrative, finance, and contracting officers.[26] Billets requiring civilian personnel to deploy vary by year, with current requirements hovering around 100–150 billets.[27] The agency's Expeditionary Readiness Center provides training, administrative, and medical support to deploying personnel. The center provides many of the same services to other Intelligence Community organizations through memoranda of understanding or agreement, including the National Geospatial-Intelligence Agency, the National Reconnaissance Office, the Office of the Director of National Intelligence, and the National Security Agency. Like the Defense Logistics Agency and

---

[24] U.S. Drug Enforcement Administration, "DEA Mission Statement," web page, undated.

[25] Interviews with Drug Enforcement Administration officials, September 24, 2014.

[26] Defense Intelligence Agency, "About DIA," web page, undated.

[27] Interviews with Defense Intelligence Agency officials, October 10, 2014.

the Defense Contract Management Agency (discussed later), the Defense Intelligence Agency is part of the DoD Fourth Estate.

Defense Logistics Agency, U.S. Department of Defense

"As America's combat logistics support agency, the Defense Logistics Agency provides the Army, Navy, Air Force, Marine Corps, other federal agencies, and combined and allied forces with the full spectrum of logistics, acquisition and technical services."[28] The Defense Logistics Agency employs approximately 27,000 personnel, of whom 1,000 are military and the rest civilian. It has personnel stationed overseas at distribution centers in support of routine missions, but the agency has also deployed up to 300 civilian staff in support of contingency operation requirements.

Defense Contract Management Agency, U.S. Department of Defense

The Defense Contract Management Agency consists of more than 11,900 civilians and military personnel who manage the execution of contracts on behalf of DoD that cover more than 20,000 contractors.[29] Although the agency was established only in 2000, it has undergone significant changes—from a primarily domestic contract oversight role to that of an expeditionary force provider. In that capacity, the agency at one time deployed up to 450 civilian contracting and support personnel, but with the subsequent drawdown of forces, current requirements range from 50 to 100 deployed personnel.[30]

UK Ministry of Defence Support-to-Operations Team

MOD's S2O office was established in 2006 to enable the generation, deployment, and subsequent redeployment of MOD civilians in support of overseas operations. Its policy and communication team is responsible for deployment policy, rules, and guidance, as well as promoting the program and managing information disseminated to

---

28 Defense Logistics Agency, "DLA at a Glance," web page, undated.

29 Defense Contract Management Agency, "About the Agency," web page, undated.

30 Interviews with Defense Contract Management Agency officials, September 11, 2014.

the S2O community.[31] The safety and security team is responsible for managing the risks associated with deploying to operational theaters and for the policies concerning safety, security, and visits. Finally, the administrative support team handles the administrative elements of deployments, including booking flights, processing operational allowances, and scheduling individuals for training. The roles that this team supports include policy advisers, civil secretaries, media advisers, and operational analysts.[32] Each role has a designated senior-level official who is responsible for maintaining adequate pools of volunteers for deployment. For some roles, this also includes high-readiness pools; however, these pools are currently in an early stage of development.

Canadian Department of National Defence

Since its involvement in Afghanistan in 2001, the Canadian Department of National Defence has deployed civilian specialists to operational theaters. Personnel deployed include medical specialists, morale and welfare staff, policy advisers, and intelligence analysts.[33] While many individuals deploy under the public service umbrella, some are sourced through the Canadian Department of Foreign Affairs, Trade and Development. The civilian requirement varies across missions but has entailed an average of 80 deployed personnel working on the ground in Afghanistan.[34]

Australian Defence Science and Technology Organisation

DSTO (now called the Defence Science and Technology Group) consists of approximately 2,300 civilian staff employed as scientists, engineers, IT specialists, and technicians.[35] It is the part of the Australian Department of Defence and supports scientific analysis and research and development. As part of its mission, the organization deploys sci-

---

[31] Interview with an S2O official, August 2014.

[32] Interview with an S2O official, August 2014.

[33] Interviews with Canadian Department of National Defence officials, September 2014.

[34] Interviews with Canadian Department of National Defence officials, September 2014.

[35] Australian Department of Defence, Defence Science and Technology Group, "About DST Group," web page, undated. In this report, we use the name of the organization at the time this research was conducted.

entists in support of military operations to provide immediate, on-the-ground advice and assistance. Personnel deployed in these roles include operational analysts, anthropologists, and cultural advisers.[36] Scientists are paired with military personnel and deploy as a team. This pairing is established during predeployment training and continues throughout the deployment. DSTO's requirement for particular civilian skill sets has varied over time, from geospatial specialists to analysts skilled in developing metrics to understand strategic impact.[37]

Civilian Planning and Conduct Capability, European External Action Service

CPCC, part of the EEAS, supports the sourcing of staff for Common Security and Defence Policy missions, among other responsibilities. The first such mission was launched in 2003. Since then, the European Union has launched 24 civilian missions and military operations.[38] In 2013, CPCC supported ten Common Security and Defence Policy civilian missions, including training missions, border and judicial system support, support for security-sector reform, support to authorities in combating terrorism and organized crime, and more general advice or assistance with defense reform. Deployed personnel come from EU member states and third-party states (those outside of the European Union).[39] The missions range in duration, depending on the mission mandate. Missions involve the deployment of roughly 3,200 military personnel, and CPCC had around 3,700 civilians deployed at the time of this research. The Civilian Headline Goal 2010 aimed to improve the European Union's civilian capability to respond effectively to crisis management tasks in the context of the Common

---

[36] Interview with a DSTO official, September 25, 2014.

[37] Interview with a DSTO official, September 25, 2014.

[38] European Parliament, Directorate General for External Policies of the Union, Policy Department, *CSDP Missions and Operations: Lessons Learned Processes*, Luxembourg: European Union Publications Office, April 2012.

[39] Interview with a CPCC official, October 10, 2014.

Security and Defence Policy.[40] One focus was on improving the capabilities and capacity of civilians, for instance through improved predeployment training. In 2011, Europe's New Training Initiative for Civilian Crisis Management (known as ENTRi) was launched to prepare and train crisis management personnel in a rigorous and standardized manner.[41]

Typology of Civilian Deployment Models

Overall, we identified four models that these organizations have applied to deploy civilians. The models differ along two main dimensions: the extent to which they sourced individuals to deploy from within the organization's existing civilian ranks (internal sourcing), as opposed to searching for candidates external to the organization (external sourcing), and the extent to which the organizations had a pool of preidentified individuals prior to the issuance of requirements (proactive sourcing), as opposed to identifying candidates for positions after requirements had been issued (reactive sourcing).[42] Table 3 categorizes the 17 analogous organizations into the four deployment sourcing models.

It is important to note that some of these organizations can be classified into more than one category, depending on the office in which it is housed. We therefore categorized each organization based on the *predominant* sourcing model it used to fill the *majority* of civilian requirements.

Reactive Internal Sourcing

We termed the first model *reactive internal sourcing*. In this model, a requirement is identified through either a top-down or bottom-up

---

[40] European Union, Civilian Capabilities Improvement Conference and General Affairs and External Relations Council, "Civilian Headline Goal 2010," November 19, 2007.

[41] Europe's New Training Initiative for Civilian Crisis Management, homepage, undated.

[42] There are benefits and drawbacks to both reactive and proactive sourcing, depending on the situation in question.

Table 3
Deployment Models, by Agency

	Reactive Sourcing	Proactive Sourcing
Internal Sourcing	U.S. Customs and Border Protection U.S. Drug Enforcement Administration U.S. Defense Intelligence Agency Canadian Department of National Defence DoS Bureau of South and Central Asian Affairs U.S. Department of Homeland Security Office of International Affairs	U.S. Defense Contract Management Agency U.S. Defense Logistics Agency FEMA DoS Bureau of Diplomatic Security UK MOD S2O Australian DSTO (now the Defence Science Technology Group)
External Sourcing	DoS Afghanistan and Pakistan Strategic Partnership Office EEAS CPCC	USAID Office of Transition Initiatives USAID CS3 DoS Bureau of Conflict and Stabilization Operations EEAS CPCC, Civilian Response Teams

process (see Figure 1).[43] The requirement triggers a recruitment process internal to the organization—for instance, within the civil service of a defense department. Individuals are selected to fill the requirement, after which they undertake any required predeployment training or preparation (e.g., medical screening and vaccinations, visas, clearances, culture training, hostile environment training) as needed for the specific deployment. Individuals deploy to their posting, and after deployment they return to the post they had occupied prior to deployment.[44]

---

[43] The requirement in some instances had to be validated through engagement between theater and home office leadership to ensure that it was valid and that a civilian was best placed to fill it before individuals were sourced (interviews with MOD S2O officials, August 2014).

[44] In some organizations, individuals were given a preview of life during operations to ensure that people's decisions to volunteer were based on realistic information about the position. In the literature on organization selection reviewed for this study, this practice is often referred to as a "realistic preview," happening prior to the application process (interviews with MOD

Figure 1
Reactive Internal Sourcing Model

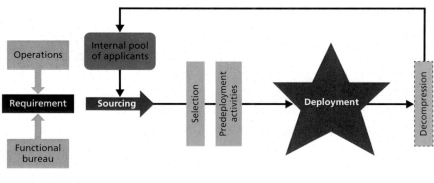

RAND *RR1249-1*

Only a few organizations specifically mentioned decompression as part of the process.[45]

Benefits and Constraints

The reactive internal sourcing model is beneficial in that it entails (and enables) longer-term organizational ownership of the skills required by deployed civilians. Because it focuses solely on candidates internal to the organization, it ensures that the organization maintains these civilian capabilities within its overall workforce following any particular deployment. Personnel who are deploying complete training and preparation just prior to deployment, such that costs are not incurred well in advance of deployment and the training can be targeted for the specific deployment. In many cases, the deployment training and additional skills that are developed by civilians while deployed have added benefits that can be applied back to the home organization upon return. The individuals involved with the candidate selection process—

S2O officials, August 2014; interview with a DSTO official, September 25, 2014; interviews with Defense Intelligence Agency, Drug Enforcement Administration, and U.S. Customs and Border Protection officials, September and October 2014).

[45] Interviews with MOD S2O officials, August 2014; interviews with a DSTO official, September 25, 2014; interviews with Canadian Department of National Defence officials, September 2014; interviews with Defense Intelligence Agency officials, October 10, 2014.

handled by management or a selection board that is familiar with the requirement—understand the organization, mission, and capabilities and are selected according to the specific requirement.[46]

However, there are also a number of potential drawbacks associated with this model. The length of the process means that it is not well suited for short-notice, urgent deployments (unless there is a speedy internal process for advertising positions and quickly recruiting staff).[47] Applicants are also already employed in other roles within their organizations, which means that when they deploy, their posts are often left open without backfill. This can lead to a loss of productivity and increased risk to the home office. Additionally, for certain skill sets, the candidate pool may be too shallow under this model, causing a capability shortage if an individual with the necessary qualifications does not apply to fill the requirement.[48] The more complex the expertise needed, the more difficult it is to find suitable candidates.[49]

Other constraints include issues with reintegrating deployed personnel back into the home office. Sometimes, individuals do not want to return to their previous role or position due to their newly acquired experience.[50] Other times, employees are penalized for deploying or there is home office animosity toward the deployed person because his or her position was gapped without backfill. Individuals have returned to find that their former jobs have been filled.

Finally, some interviewees identified issues surrounding traditional human resources (HR) functions. Several challenges arose concerning the identification and selection of potential deployed person-

---

[46] Interviews with Defense Intelligence Agency officials, October 10, 2014.

[47] Interviews with CPCC officials, October 10, 2014.

[48] For instance, since 2006, there have been 43 calls for contributions to Common Security and Defence Policy civilian missions, with a goal of filling 150 posts. Officials reported that 109 of these posts were filled through this process, though only 23 were filled with individuals from the expert pool (interviews with CPCC officials, October 10, 2014). For selected organizations that have maintained a high deployment tempo over multiple years, the candidate pool is sufficiently deep, but candidate availability has been diminished by consecutive deployments.

[49] Interviews with CPCC officials, October 10, 2014.

[50] Interviews with MOD S2O officials, August 2014.

nel, performance evaluations, and the overall flow of information from the HR office to potential volunteers. Other interviewees raised the issue of unfair promotion practices.[51]

Proactive Internal Sourcing

The second model identified through our analysis is the *proactive internal sourcing model*. Similar to reactive internal sourcing, under this model, deployed civilians are sourced from within an organization's existing civilian employee pool. However, rather than waiting for a specific requirement to be identified, the most probable requirements to emerge are preidentified. Civilians within the organizations can then apply to be part of a *readiness pool* that will be used to source the set of emergent requirements. Individuals are identified per internal selection processes and undergo required predeployment training and screening. Upon completion, individuals are placed in the readiness pool, as shown in Figure 2.

Figure 2
Proactive Internal Sourcing Model

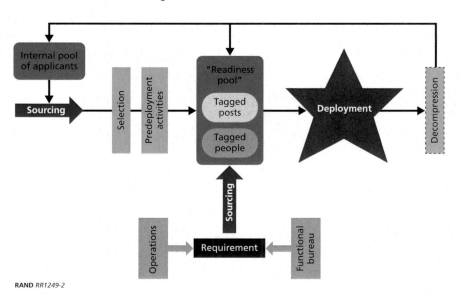

RAND RR1249-2

---

[51] Interview with Defense Intelligence Agency, Drug Enforcement Administration, and U.S. Customs and Border Protection officials, September and October 2014.

Benefits and Constraints

The primary benefit of the proactive internal sourcing model is the ability to deploy on relatively short notice, because individuals are pre-selected and prepared for a set of likely missions. These individuals are then put into a readiness pool where they continue to function in their normal capacity until notified of an upcoming deployment. Entrance into the readiness pool typically happens through one or two methods. First, individuals are hired into the organization in a "tagged post," in which an offer of employment is conditional upon agreement to be called to deploy. Second, an individual with a certain skill set that is valuable to the organization in both deployed and nondeployed environments are identified as "tagged people" and are also placed into the readiness pool.

Similarly to the previous model, ownership of the skills needed for deployments are retained within the organization. Furthermore, there is an opportunity in both internally sourced models for organizations to learn from the experience and retain the expertise of deployed personnel upon their return. As with the previous model, drawbacks associated with proactive internal sourcing include a lack of backfill for (and the requirement to hold open) home office postings, as well as related difficulties as the civilian attempts to reintegrate into the home office after deployment. Individuals in the readiness pool are not guaranteed to deploy. For example, the requirement might never emerge, it might take too long for the requirement to emerge, or there could be an issue with retention of personnel within the pool.[52] Multiple organizations that utilize this model encourage HR staff or the office responsible for deploying personnel to actively monitor the pool to ensure that it is appropriately sized to meet requirements and that personnel are not in the pool so long that they lose interest in deploying.[53] The cost of predeployment training is also incurred regardless of deployment, and civilians in the readiness pool may need refresher training or new training, depending on how much time has elapsed since their recruitment into the pool.

---

[52] Interview with a DSTO official, September 24, 2014.

[53] Interviews with MOD S2O officials, August 26, 2014.

A final drawback of the proactive internal sourcing approach is that the forecasted requirement against which the individuals were originally recruited may evolve and differ from actual future requirements.[54] Therefore, some process must be established to routinely validate the set of requirements and the appropriateness of the skill sets represented in the readiness pool. For example, some agencies have a quarterly validation panel that looks at current and future requirements; the readiness pool is subsequently adjusted according to these new requirements.[55]

Reactive External Sourcing

The third model identified in this comparative analysis is the *reactive external sourcing model*. In this model, individuals are drawn from external sources for deployable civilian positions. That is, a specific requirement is identified, the organization advertises the requirement externally, and an outside expert is hired to fill the requirement, as shown in Figure 3. The organization usually covers the costs associated with any necessary training, medical screening, visas, and security clearances for the individual in question. Following deployment, employees hired under this model are no longer affiliated with the organization, often returning to their former posts with other organizations (including universities). The Afghanistan and Pakistan Strategic Partnership Office in DoS is one organization that utilizes this approach.[56] We also categorized CPCC within this category, though the distinction between internal and external sourcing in CPCC is less clear-cut because a call for contributions is sent out to member states, which then look internally to their government departments for candidates. Once a candidate is selected, CPCC will fill the requirement and the candidate will deploy for the mission. If a call for contributions has

---

[54] Interviews with Canadian Department of National Defence officials, September 2014; interviews with CPCC officials, October 10, 2014.

[55] Interviews with analogous civilian deployment organizations, July–November 2014; interviews with U.S. government officials, 2014.

[56] Interview with an Afghanistan and Pakistan Strategic Partnership Office official, September 10, 2014.

Figure 3
Reactive External Sourcing Model

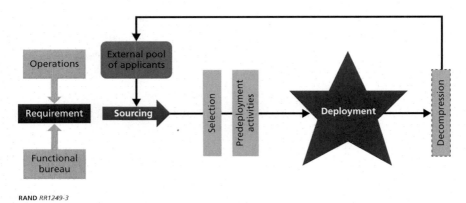

RAND *RR1249-3*

been sent out twice without enough volunteers, the call is expanded to non-member states.

Benefits and Constraints

The benefits of the reactive external sourcing model are similar to those of the reactive internal sourcing model, particularly with regard to targeting predeployment training only to those who will be deployed. The selection procedure is focused on finding the most highly qualified individuals matching the requirement. Because individuals are sourced externally, there is no issue with gapping home office assignments within the organization, and the costs are not incurred unless individuals are actually hired to fulfill a requirement and deploy.

Challenges associated with this model include an ever-present question as to whether the skills needed for any particular requirement will be readily available in the external environment. In our interviews, this was not typically a concern.[57] However, it is possible to have a scenario in which specific requirements are hard to fill because the capability is not readily available outside of the organization. An example

[57] A few interviewees mentioned difficulty finding personnel with the necessary skill sets for highly technical work, such as electricians, rule-of-law specialists, DNA analysts, air traffic controllers, and English-language specialists (interview with a U.S. government official, September 2014; interviews with CPCC officials, October 10, 2014).

could be a technical expert or senior Foreign Service officer. Furthermore, if skills attractive to the home office are developed on deployment, they are not retained after deployment, as individuals return to their predeployment work status outside the organization. Often, external recruitment is a lengthy process and does not lend itself to urgent, short-notice deployments because of the U.S. Office of Personnel Management's competitive hiring authorities.[58] Finally, there may be additional training or security requirements associated with deploying external candidates that need to be considered.

Proactive External Sourcing

The fourth model identified through our analysis is the *proactive external sourcing model*. Much like the reactive external sourcing model, personnel from outside an organization are identified to fill requirements. The organization uses various planning models to forecast a set of future requirements and then hires personnel to source those requirements. In anticipation of a requirement for civilian deployment, organizations such as the Office of Transition Initiatives and CS3 set up a *bullpen*, a readiness pool of external selectees that is used to fill requirements when needed, as shown in Figure 4.[59] A practice commonly seen in organizations utilizing this model involves selectively hiring experts prior to the issuance of actual requirements, conducting predeployment training and medical screening, obtaining passports and security clearances, and then placing candidates in the bullpen, where they will wait to be called for a deployment. While in the bullpen, individuals are not paid, nor are they provided benefits. When a requirement is issued that matches the qualifications of a particular individual in the bullpen, he or she is notified for a deployment and then his or her pay and benefits are activated.

---

[58] Multiple analogous civilian organizations mentioned that the fair hiring practices mandated by the U.S. Office of Personnel Management require a thorough screening of all applicants—a process that lengthens the hiring process. See Office of Personnel Management, "Hiring Authorities: Competitive Hiring," web page, undated.

[59] Interviews with Office of Transition Initiatives and CS3 officials, September 2014.

Figure 4
Proactive External Sourcing Model

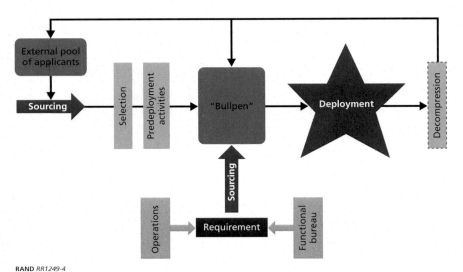

RAND *RR1249-4*

Benefits and Constraints

The period between the identified requirement and the actual deployment is likely to be shorter in the proactive external sourcing model because individuals are already preidentified, have been selected, and have undertaken required predeployment readiness preparation. Salary-related costs are incurred only after the individual deploys under this model and, depending on the organization, would be paid by either the home office or the field office. For example, interviewees reported that most salary costs for deployed personnel at one organization were paid by the home office. Other interviewees reported that the salary costs at another organization that uses a bullpen were predominately covered by the embassy that the deployed personnel supported.[60]

Most organizations do not have a need to backfill posts due to the nature and function of the bullpen. Requirements are forecasted such that the necessary qualifications are understood in general terms, and experts with the necessary knowledge and skill sets to meet these quali-

_____

[60] Interviews with officials from analogous civilian deployment organizations, July–November 2014; interviews with U.S. government officials, 2014.

fications are selected for the bullpen. Many organizations that use this model have stringent authority over the hiring and firing of personnel, and there are few bureaucratic processes associated with relieving personnel who are not a good fit.[61]

Yet, this model also has its share of drawbacks. Although salaries are paid only upon actual deployment, the organization may incur the costs associated with predeployment training up front regardless of whether an individual deploys. If the requirement changes from what was originally anticipated, there may be issues with not having the required capability or skill set in the bullpen, and the training and readiness costs are sunk expenses that cannot be retrieved for personnel who no longer meet the qualifications of evolving forecasted requirements. Finally, as seen in the previous model, the skills developed by the individual during his or her deployment are not retained easily within the organization.

Lessons from the Four Deployment Models

The four models highlight the differences in how organizations handle civilian deployments. We believe that organizations can draw on a combination of these models, emphasizing aspects that fit their specific situation and best position the organization to respond to requirements. For example, it was clear that most organizations with a requirement to deploy civilians on relatively short notice to a hostile environment chose to develop a kind of cadre; they had a process for preselecting people who could fill requirements that arose quickly.[62] However, we found that the time it took to fill a requirement for a particular civilian deployment varied greatly. A number of factors affected the speed of deployment, including the organization from which the individu-

---

[61] Interviews with officials from analogous civilian deployment organizations, July–November 2014; interviews with U.S. government officials, 2014.

[62] Interview with a FEMA official, September 19, 2014; interview with a CS3 official, September 15, 2014; interview with MOD S2O officials, August 2014; interview with a DSTO official, September 25, 2014.

als were sourced, the extent to which the required skill set was readily available, and the generic selection procedures applied within the organization.

Yet, we found it notable that, of the four models we identified, those that involved a proactive sourcing approach allowed organizations to deploy personnel significantly faster than those that involved recruiting qualified personnel after a requirement had been issued.[63] Meanwhile, organizations that relied on external sourcing models spoke to the numerous rules and regulations faced by government agencies. At times, lengthy justifications were needed to select one individual over another to adhere to fair hiring practices.[64] This indicates that, when reactive sourcing is necessary, sourcing officials may need direct or expedited hiring authorities to enhance their ability to source positions quickly. Furthermore, regardless of whether individuals were sourced internally or externally, some type of oversight organization was necessary to ensure the successful deployment of civilians.

Across the analogous organizations, we identified opportunities to pool and share existing capabilities for civilian deployments. For instance, certain agencies within the Intelligence Community share predeployment training and medical facilities, such as the Defense Intelligence Agency's Expeditionary Readiness Center.[65] Instead of maintaining individual deployment divisions, organizations could pool those resources, and one agency could provide training on behalf of the others.

There are a number of decisions to be made with regard to the size and scope of a deployable civilian capability. Planning and forecasting will help optimize the timelines associated with deployment. For instance, if the requirement is not urgent, the organization has time to use a reactive sourcing model. Although we did not directly assess the

---

[63] Experts from the CPCC Civilian Response Team pool have been deployed within five days (interviews with CPCC officials, October 10, 2014; interviews with U.S. government officials, 2014).

[64] Interview with a CPCC official, October 10, 2014; interview with an Afghanistan and Pakistan Strategic Partnership Office official, September 10, 2014.

[65] Interviews with Defense Intelligence Agency officials, September 2014.

difference in costs between sourcing external candidates versus internal candidates, it is likely that costs will differ and that the cost itself will be a factor in choosing a sourcing model.[66] Furthermore, if a skill set is required within the organization in the future, it is valuable to maintain and sustain the skill set internally rather than externally. However, if a permanent civilian workforce is built to include all necessary expeditionary civilian capabilities internally, the nature of the entire workforce may change. For some skills, this may be critical if a future surge is required. Other skills, however, may not need to be retained internally in the organization (for example, Ebola/infectious disease specialist physicians). Yet, overutilization of expeditionary civilian personnel sourced from outside the organization will entail challenges in postdeployment tracking, and the skill sets will not be readily available in the future. Each organization must decide which capabilities to retain internally and which to look for outside the organization. If the ability to quickly deploy personnel is of primary concern, preidentified personnel are recommended. To source and deploy civilians rapidly, our analysis suggests a need to closely examine the speed of the recruitment processes, the possibility of developing a preselected pool, and the possibility of making deployment part of the job description. To that end, the establishment of a preidentified readiness pool will require the accurate forecasting of future requirements and likely mission sets.

Related to this point, the organizations analyzed considered the positions that they were looking to fill with civilians; they also scrutinized the requirements for civilian deployment to ensure that the post was required and that only a civilian could fill it.[67] For instance, the Canadian Department of National Defence, Australian DSTO, UK MOD, and EU CPCC all draw their civilian deployees from volunteers—that is, individuals deploy on a voluntary basis. Interviewees from these organizations noted that, within their workforces, they

---

[66] Such a comparative cost analysis assessing the relative expense of each of the four deployment models outlined here would be a fruitful area for future research.

[67] Interviews with MOD S2O officials, August 2014; interviews with Canadian Department of National Defence officials, September 25, 2014.

had capable individuals who were interested in volunteering, and they found that very few people withdrew their offer to deploy.[68]

Meanwhile, among the U.S. agencies examined here, some mandated that specific individuals deploy (Bureau of Diplomatic Security, Bureau of Conflict and Stabilization, Office of Transition Initiatives, Defense Logistics Agency),[69] others requested volunteers (Office of International Affairs, U.S. Customs and Border Protection, Drug Enforcement Agency),[70] and others used a combination of factors to make this decision, including the source of the requirement, the time needed to fill the position, and whether the skill set was internal to the organization or had to be found externally. Organizations within the DoD Fourth Estate typically had a mixed civilian/military workforce. In organizations with a well-defined set of requirements to deploy personnel, there were usually systems in place to facilitate individuals volunteering to fill a requirement. In sum, most of the organizations examined had well-defined policies that clearly articulated duties and procedures surrounding the deployment process. However, other agencies lacked many basic policy documents and consistently handled their deployment procedures on an ad hoc basis.[71]

Conclusions and Recommendations

This research points to a number of interesting findings regarding the benefit of a long-term strategy aimed at developing a viable civilian deployment practice that will be sustainable and the specific practices and processes that organizations might usefully employ on a day-to-

---

[68] Various interviews with officials from analogous organizations, May–November, 2014.

[69] Interview with Bureau of Diplomatic Security officials, August, 14, 2014; interview with CSO, September 18, 2014; interviews with Office of Transition Initiatives officials, September 15, 2014; interviews with Defense Logistics Agency officials, July 19, 2014.

[70] Interview with an Office of International Affairs official, August 25, 2014; interview with U.S. Customs and Border Protection officials, September 18, 2014; interviews with Drug Enforcement Administration officials, September 24, 2014.

[71] Various interviews with officials from analogous organizations, May–November 2014.

day basis to ensure effective, efficient civilian deployments. Notably, the findings at both levels are intertwined, with the strategic findings being necessary for the support and establishment of the day-to-day practices. It is useful to consider the management of civilian deployment capabilities as being divided into three categories of activities: policy, planning and strategy, and operations. *Policy* responsibilities entail writing policy to determine the guidelines for civilian deployments. *Planning and strategy* responsibilities entail mission-based, scenario-specific forecasting and strategic human capital planning. *Operational* responsibilities entail the assignment of requirements, sourcing, readiness preparation, and during/post deployment tracking of expeditionary civilians. Figure 5 illustrates these distinctions.

Policy
Championing Expeditionary Civilian Capabilities
The military may be unaware of the benefits of drawing on civilian capabilities for deployed missions, and organizations must make a conscious effort to market civilian capabilities in this regard. This is a relatively day-to-day practice that organizations can employ to ensure that defense leaders and other "customers" of deployable civilian capabilities

Figure 5
Ownership of Relevant Aspects of Civilian Deployment

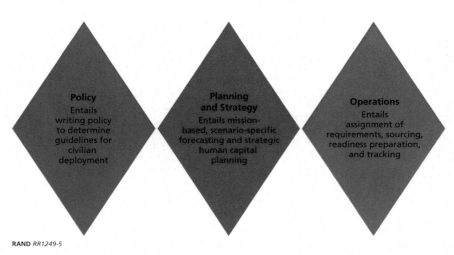

Policy
Entails writing policy to determine guidelines for civilian deployment

Planning and Strategy
Entails mission-based, scenario-specific forecasting and strategic human capital planning

Operations
Entails assignment of requirements, sourcing, readiness preparation, and tracking

are aware of the benefits that deployed civilians can bring to an operation. Marketing serves two primary functions. First, it provides those unfamiliar with civilian deployment a better understanding of what it means to be a deployed civilian. This gives applicants a realistic preview of the roles in which they could possibly deploy and helps adjust expectations for both those generating requirements and the civilians who consider volunteering for a deployment. Second, marketing ensures that individuals who request deployed civilians understand both the current and future capabilities that are available. This helps those generating requirements understand where deployed civilians would be most appropriately assigned and how they could be employed. Such initiatives, if supported over time with evidence that the organization in question can effectively deliver the civilian capabilities marketed in a timely manner, thus promote a cultural shift in overcoming potential misconceptions about the availability, deployment process, and usability of civilians in support of military operations.

Balancing Readiness with Cost

Organizations must craft their own structure for deploying civilians, tailored to their specific needs. The speed of recruitment and the cost of readiness preparation and maintenance must be balanced against the cost of those capabilities. Those capabilities vary to different degrees under different circumstances, and some skills should likely be maintained internally over the long term (while others need not be). To that end, we conceived of a tiered approach to combine internal, external, reactive, and proactive hiring mechanisms to ensure readiness and flexibility in the force across a spectrum of possible future contingencies. Figure 6 highlights some of the possible considerations for each organization. There is not one model that is right for every organization, and each should weigh its own requirements to deploy civilians and adopt an appropriate approach.

Planning and Strategy
Speed Versus Capability

Our analysis suggests that there is a need to examine the speed of the recruitment process, the cost of maintaining deployable civilians at a

Figure 6
A Tiered Approach to Balance Readiness Needs with Cost Constraints

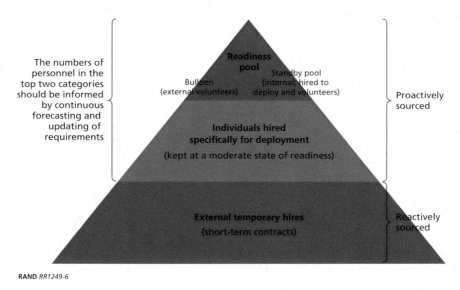

The numbers of personnel in the top two categories should be informed by continuous forecasting and updating of requirements

Readiness pool

Bullpen (external volunteers)   Standby pool (internal, hired to deploy and volunteers)

Individuals hired specifically for deployment

(kept at a moderate state of readiness)

External temporary hires

(short-term contracts)

Proactively sourced

Reactively sourced

RAND RR1249-6

given level of readiness, and the possibility of developing a preselected pool or making deployment part of the job description. Ultimately, the speed of recruitment and cost of readiness preparation and maintenance will matter to different degrees in different circumstances. Clearly, the mission type and the urgency of the deployment will matter, as will the financial resources available to support the deployment process. Furthermore, we found that if a specific skill set is required within an organization for the future, it is valuable to develop and sustain the capability internally rather than externally to be able to deploy individuals with that capability in the future.

Planning and Forecasting
The extent to which planning actually incorporates considerations of expeditionary civilian requirements is questionable at this point, at least as indicated by our interviews. A failure to effectively integrate expeditionary civilians into planning for various scenarios and missions hinders the development of realistic expectations for the numbers of expeditionary civilians required in any given situation. It also ultimately

poses a challenge to the integration of these civilians by decreasing organizations' ability to plan ahead for backfill needs when one of their civilians deploys. Therefore, organizations need to dedicate resources to the planning and forecasting of future requirements to optimize the organizational design associated with civilian deployments. Figure 7 shows an example of such an approach.

Operations
Centralized Versus Decentralized
Organizations must determine the level at which to manage many of the processes that govern civilian deployments. Should capability and oversight be retained at a headquarters level, or should it be pushed down to the operational agency that deploys personnel? The extent to which an organization centralizes that authority determines the speed, budget, capability, and resources that it requires. A decentral-

Figure 7
Scenario-Based, Mission-Specific Forecasting for the
Long-Term Utilization of Deployable Civilians

RAND *RR1249-7*

ized recruitment, screening, and selection approach does not necessarily work when an organization must meet several different deployable position requirements. A centralized process may be slow to respond to emergent requirements.

A good example showing the difference between centralized and decentralized process is readiness preparation. It involves training and the processing of any necessary clearances (e.g., medical, security), as well as the provision of visas, passports, and other administrative documentation necessary for deployment. While the headquarters are best resourced to manage readiness preparation for civilian employees tasked with a mission, they may not be prepared to manage readiness preparation for civilians who will conduct missions for another organization or agency.

The tracking of civilians both pre- and postdeployment is another critical function that can highlight the differences of centralized and decentralized approaches. Tracking entails maintaining contact with a deployed civilian both during deployment and for a period of time postdeployment to assist with administrative, HR, or occupational issues and to screen the individual for any deployment-related health problems following his or her return. We recommend that a centralized headquarters-level organization be tasked with oversight and management of the assignment of requirements and the recruitment, screening, and selection of expeditionary civilian candidates.

Sharing Resources

We identified opportunities to pool and share existing enabling capabilities for civilian deployments, including the sharing of training facilities. Consolidating the responsibility to deploy personnel from multiple organizations to one deployment center saves on overhead, personnel, and operational costs.[72] There are several options that could be utilized to deploy civilians through one center. Centers could be established

---

[72] There is one center currently in operation in the DoD that deploys civilians from other agencies. Funding for this center comes primarily through one agency, which typically provides access to the center for other agencies through a memorandum of agreement. This memorandum typically addresses overall responsibilities for each agency, the type of training to be provided, and the transfer of funds between agencies.

based on geographic location or on a specific function. For instance, organizations that routinely deploy personnel to the same geographic location could conduct joint predeployment training, ensure that the same medical procedures are followed, and coordinate on visas, logistics, and other supported activities.

Areas for Future Research

We found a dearth of studies tallying the costs of different deployment models or, indeed, the costs associated with different utilization of parts of the whole force. A potential avenue for further research could include the development of a costing framework for each of the models that would allow transparency of costs associated with both the deployment process as a whole and different elements of each of the models.

While we found that improved planning and forecasting of requirements for deployable civilian capabilities could be useful in informing civilian deployment processes, it was not within the scope of this study to explore how best to pursue such forecasting. Future research on this topic could therefore usefully examine, compare, and analyze various methods for forecasting deployable civilian requirements and determine the method or methods most likely to accurately forecast such requirements. Other possible research could focus on practical aspects of civilian deployment. For example, safety considerations, operating with military personnel, performance metrics, pre- and postdeployment stress evaluations, and the psychological impact of deployments could all be explored.

Finally, research could be conducted to further elaborate and refine the deployment models. For instance, future analysis could test additional HR management elements (e.g., different recruitment practices, different decompression practices, or different levels of predeployment training) with the models and evaluate their impact on deployment outcomes, such as the speed of deployment, quality of performance, or the cost of the deployment process.

References

Australian Department of Defence, Defence Science and Technology Group, "About DST Group," web page, undated. As of December 23, 2015:
http://www.dsto.defence.gov.au/discover-dsto/about-dst-group

Defense Contract Management Agency, "About the Agency," web page, undated. As of December 23, 2015:
http://www.dcma.mil/about.cfm

Defense Intelligence Agency, "About DIA," web page, undated. As of December 23, 2015:
http://www.dia.mil/About.aspx

Defense Logistics Agency, "DLA at a Glance," web page, undated. As of December 23, 2015:
http://www.dla.mil/AtaGlance.aspx

Dunigan, Molly, Susan S. Everingham, Todd Nichols, Michael Schwille, and Susanne Sondergaard, *Expeditionary Civilians: Creating a Viable Practice of Department of Defense Civilian Deployment*, Santa Monica, Calif.: RAND Corporation, RR-975-OSD, 2016. As of May 2016:
http://www.rand.org/pubs/research_reports/RR975.html

European Parliament, Directorate General for External Policies of the Union, Policy Department, *CSDP Missions and Operations: Lessons Learned Processes*, Luxembourg: European Union Publications Office, April 2012. As of December 23, 2015:
http://www.tepsa.eu/download/CSDP%20Missions%20and%20Operations-%20Lessons%20Learned%20Processes%20(DG-%20External%20Policies).pdf

European Union, Civilian Capabilities Improvement Conference and General Affairs and External Relations Council, "Civilian Headline Goal 2010," November 19, 2007. As of December 23, 2015:
http://www.consilium.europa.eu/uedocs/cmsUpload/Civilian_Headline_Goal_2010.pdf

Europe's New Training Initiative for Civilian Crisis Management, homepage, undated. As of December 23, 2015:
http://www.entriforccm.eu

"Memorandum of Agreement Between the Departments of State and Defense on the Protection and Evacuation of U.S. Citizens and Nationals and Designated Other Persons from Threatened Areas Overseas," September 1997. As of December 23, 2015:
http://prhome.defense.gov/Portals/52/Documents/PR%20Docs/DOS-DOD%20 Memo%20of%20Agreement%20on%20Protection%20and%20Evacuation.pdf

Office of Personnel Management, "Hiring Authorities: Competitive Hiring," web page, undated. As of December 23, 2015:
http://www.opm.gov/policy-data-oversight/hiring-authorities/competitive-hiring

Office of the President of the United States, *National Security Strategy*, Washington, D.C., May 2010. As of December 23, 2015:
http://www.whitehouse.gov/sites/default/files/rss_viewer/national_security_ strategy.pdf

U.S. Customs and Border Protection, "About CBP," web page, undated. As of December 23, 2015:
http://www.cbp.gov/about

U.S. Department of Defense, *Quadrennial Defense Review Report*, Washington, D.C., February 6, 2006.

————, *Sustaining U.S. Global Leadership: Priorities for 21st Century Defense*, Washington, D.C., January 2012.

————, *Quadrennial Defense Review Report 2014*, Washington, D.C., March 4, 2014.

U.S. Department of Defense Directive 1404.10, *DoD Civilian Expeditionary Workforce*, January 23, 2009.

U.S. Department of State, *Post Management Organization, Chief of Mission Authority and Overseas Staffing*, Foreign Affairs Manual 2 FAH-2, H-110, July 18, 2014. As of December 23, 2015:
http://www.state.gov/documents/organization/89604.pdf

U.S. Department of State, Bureau of Conflict and Stabilization Operations, web page, undated. As of December 23, 2015:
http://www.state.gov/j/cso

U.S. Department of State, Bureau of Diplomatic Security, web page, undated. As of December 23, 2015:
http://www.state.gov/m/ds

U.S. Drug Enforcement Administration, "DEA Mission Statement," web page, undated. As of December 23, 2015:
http://www.dea.gov/about/mission.shtml

U.S. Joint Chiefs of Staff, *The National Military Strategy of the United States of America: Redefining America's Military Leadership*, Washington, D.C., February 2011.

———, *Department of Defense Dictionary of Military and Associated Terms*, Joint Publication 1-02, Washington, D.C., as amended through November 15, 2015.